Organic Chemistry Workbook:
Additional Problems to Accompany
Vollhardt and Schore's
Organic Chemistry
Fifth Edition

Ahamindra Jain
Harvard University

W. H. Freeman and Company
New York

© 2008 by W.H. Freeman and Company

All rights reserved.

Printed in the United States of America

ISBN-13: 978-1-4292-0247-5
ISBN-10: 1-4292-0247-5

First Printing

W. H. Freeman and Company
41 Madison Avenue
New York, NY 10010
Houndmills, Basingstoke
RG21 6XS, England
www.whfreeman.com

Chapter 1 Worksheet A

Draw the TWO BEST Lewis Structures for OZONE, [O_3], by completing the templates provided (8 pts). Retain the geometry that is drawn in the template, and include any lone pairs and formal charges on the atoms in each structure. Depict the π-system of ozone in a 3D representation in the plane provided (6 pts), with p-orbitals perpendicular to the plane.

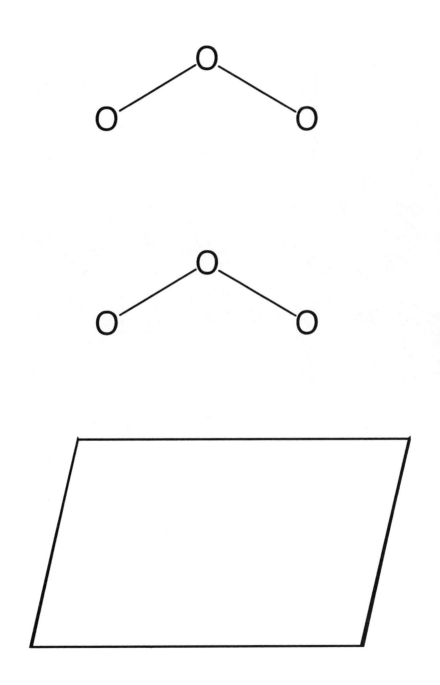

Chapter 1 Worksheet B

Complete the following Lewis Structures, and draw one additional resonance form for each. Indicate the formal charge and hybridization at each atom. PLEASE DO NOT ALTER THE GEOMETRY OF THE STRUCTURE SHOWN (12 pts).

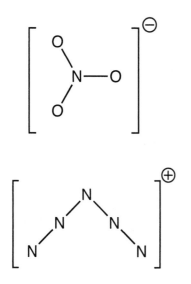

Name:_____   Date:_____

Chapter 1 Worksheet C

The molecular formula NO_2 can correspond to several different molecules, with various net charges and geometries. Assign formal charges to each of the atoms in the structures below, and an overall charge to each molecule (10 pts). Circle the two resonance forms (5 pts).

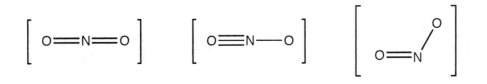

Chapter 1 Worksheet D

a) Provide two valid Lewis Structures for the formyl cation, $[HCO]^{+}$, by completing the templates provided. Retain the geometry that is drawn in the template, and include any lone pairs and formal charges on the atoms in each structure (5 pts). Circle the best resonance form, and briefly explain why (5 pts).

b) The C-C bond of ethane spins freely, since there is no change in the σ bond overlap on rotation. In the first plane provided, draw ethene (C_2H_4), showing π bond overlap (the σ bonds are in the plane, 6 pts). In the second plane provided, draw twisted ethene, where the *p*-orbitals on each carbon do not overlap (4 pts).

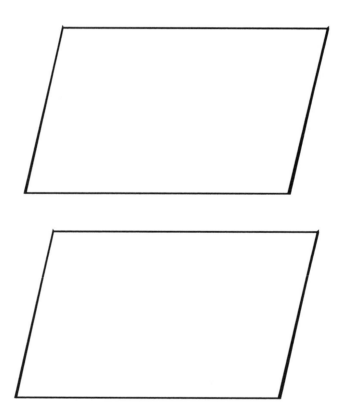

Name:_____   Date:_____

Chapter 1 Worksheet E

a) Draw the best Lewis Structure for $[H_3COCH_2]^+$ (4 pts). Be sure to clearly indicate geometry at both carbons and at oxygen (6 pts).

b) In the plane provided below, draw a clear, 3-D representation of the best Lewis Structure for $[H_3COCH_2]^+$, including bond angles, and including any orbitals that are involved in the "unusual stabilization" of this cation (any π-type orbitals should be perpendicular to the plane, 8 pts). Draw an arrow to indicate any C-O bonds that are freely rotating at room temperature (2 pts).

Name:_____   Date:_____

Chapter 2 Worksheet A

COMPLETE the arrow to indicate the direction in which each of the following equilibria are favored (9 pts). Based on your knowledge of pK_as, calculate $\Delta G°$ at 298 K for any ONE of the reactions (6 pts).

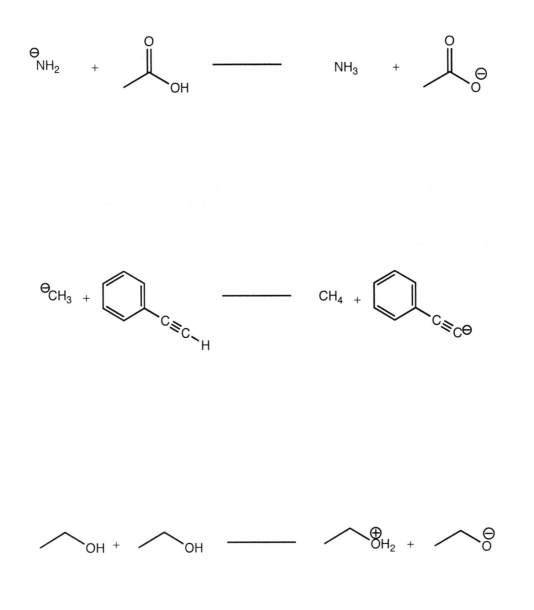

Name:_____   Date:_____

Chapter 2 Worksheet B

The following reactions are often invoked in mechanisms in organic chemistry **INCORRECTLY**. Use an arrow to indicate the direction in which the equilibria are favored (9 pts). Based on your knowledge of pK_as and by using the Gibbs Equation ($\Delta G° = -R{\cdot}T{\cdot}\ln(K_{eq})$), calculate $\Delta G°$ for ANY ONE of the three reactions (5 pts).

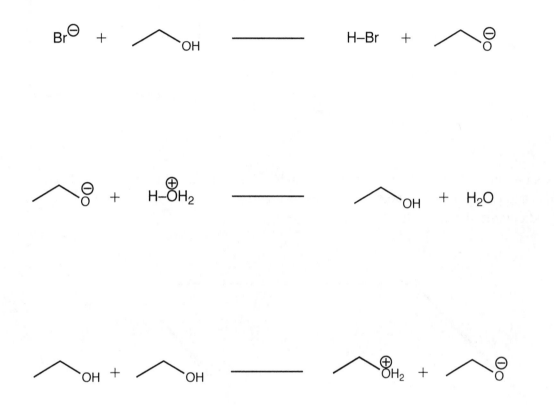

Name:_____   Date:_____

Chapter 2 Worksheet C

Nitromethane is quite acidic, due to the stability of the nitromethane anion. In the box provided, list the two factors that afford stabilization of nitromethane anion (4 pts).

$$H_2\overset{\ominus}{C}-NO_2$$

1.

2.

Draw an accurate 3-D representation of nitromethane anion, including any delocalized orbitals and the geometry at both carbon and nitrogen (please draw any π-type orbitals perpendicular to the plane, 6 pts).

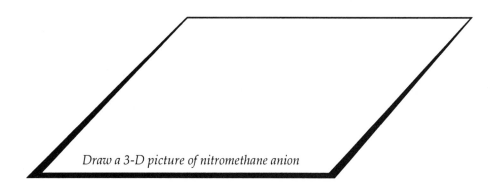

Draw a 3-D picture of nitromethane anion

Chapter 2 Worksheet D

Malonic acid has pK_as of 2.9 and 5.7 for the first and second carboxylic acid protons. Use a few words in each of the boxes to explain why these pK_as are different from the pK_a of acetic acid (8 pts), and calculate the energetic cost of each of these pK_a-perturbations (4 pts) (WITHOUT THE GIBBS EQUATION).

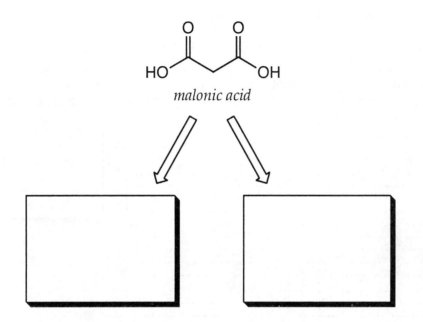

malonic acid

Name:_____   Date:_____

Chapter 2 Worksheet E

Using the templates below, draw Newman projections of the three stable conformers of 2,3-dimethylbutane looking down the bond between C_1 and C_2 as labeled below (3 pts). Complete the Newman projections of the eclipsed conformations beneath the appropriate dihedral angles (3 pts). Beneath each structure, TABULATE the NUMBER of unfavorable interactions in each conformation (gauche interactions, CH/CH eclipsing, CH/CCH$_3$ eclipsing, and CCH$_3$/CCH$_3$ eclipsing interactions (9 pts).

There are THREE stable conformations, A, B, and C. The barrier to interconversion from B to A or C is 14 kJ/mol, and from A to C is 15 kJ/mol. DEDUCE the energetic cost of each type of interaction in this system (6 pts, SHOW YOUR WORK). Draw a QUANTITATIVE plot of energy versus CCCC dihedral angle (4 pts, no partial credit).

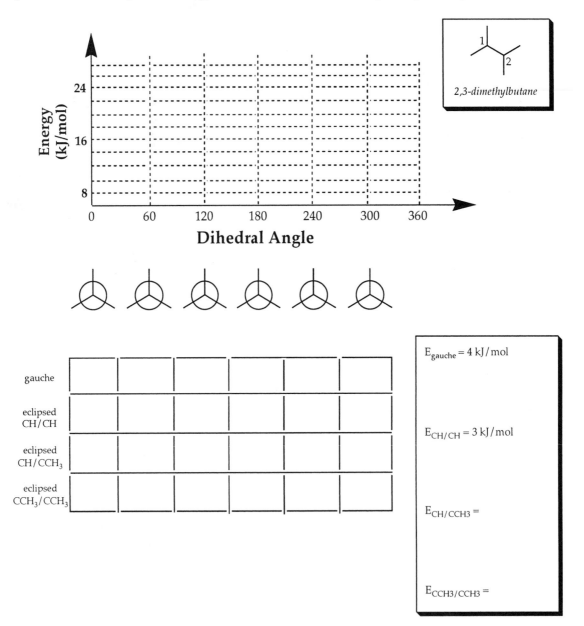

2,3-dimethylbutane

$E_{gauche} = 4\ kJ/mol$

$E_{CH/CH} = 3\ kJ/mol$

$E_{CH/CCH3} =$

$E_{CCH3/CCH3} =$

Name:_____   Date:_____

Chapter 3 Worksheet A

Chlorine radical, which is readily formed by the irradiation of chlorine gas, reacts with 2-methylbutane to form t-butyl chloride and 1-chloro-2-methylbutane, in a 1:1.9 ratio.

a) Noting that any of 9 H's may be abstracted to form 1-chloro-2-methylbutane, and that only 1 H can be abstracted to form *t*-butyl chloride, how much more reactive is the C-H than the CH_3 hydrogens (4 pts)?

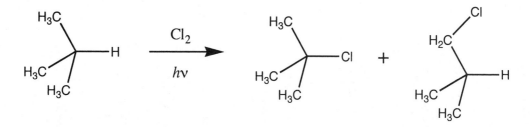

b) Which bond is weaker, the C-H or the CH_2-H bond? Explain the difference in these BDEs (4 pts).

c) Using proper CURVED-ARROW notation, write a mechanism for the formation of *t*-butyl chloride, from 2-methylbutane, chlorine radical, and Cl_2 (6 pts).

Name:_____   Date:_____

Chapter 3 Worksheet B

Carry out radical chlorination of the hexane isomer below to obtain all the isomers of $C_6H_{13}Cl$ (8 pts). Count the number of distinct hydrogens in each isomer (12 pts).

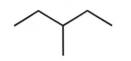

Name:_____   Date:_____

Chapter 3 Worksheet C

Provide a DETAILED, ARROW-PUSHING MECHANISM for the transformation below (8 pts).

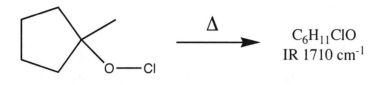

$$\xrightarrow{\Delta}$$

$C_6H_{11}ClO$
IR 1710 cm^{-1}

Chapter 3 Worksheet D

The O-Cl bond below is readily cleaved by light, to afford two radicals. No other species are present in the solution. Provide a detailed, arrow-pushing mechanism for the reaction shown (9 pts).

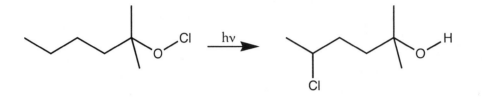

Chapter 3 Worksheet E

Spiropentane, shown on the left below, has no double bonds, but reacts with Cl_2 in the presence of light to give the two products shown. Given that Cl_2 is readily cleaved by light, provide an arrow-pushing mechanism for the formation of the common intermediate shown (3 pts), and of each product, from the common intermediate (6 pts). What structural feature of spiropentane enables it to react with Cl• in this manner (3 pts)?

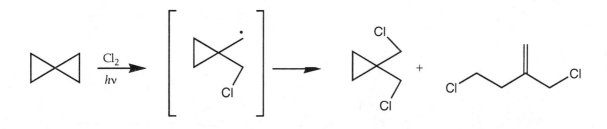

Name:_____   Date:_____

Chapter 4 Worksheet A

Draw a line to connect the structures in the first and second column that experience similar unfavorable energy corrections (6 pts). Name any TWO of these corrections on the appropriate line on the right (4 pts).

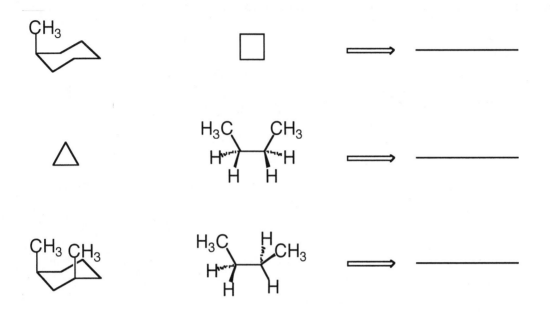

Name:_____   Date:_____

Chapter 4 Worksheet B

Match the equilibrium below to the appropriate $\Delta G°$ in the column on the right (12 pts).

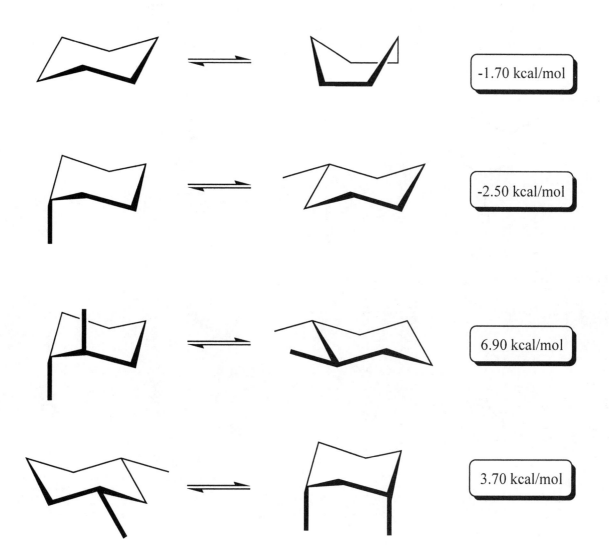

Chapter 4 Worksheet C

Draw the two interconverting conformers of the cyclohexane derivative below, in **proper chair representations** (4 pts). Given that $\Delta G°_{EQ \to AX}$ for F is 1.0 kJ/mol, and that $\Delta G°_{EQ \to AX}$ for Ph is 13 kJ/mol, calculate $\Delta G°$ for the equilibrium between the conformers (2 pts), and circle the more stable isomer (1 pt). What is K_{EQ} for this equilibrium (3 pts)?

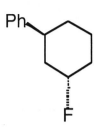

Chapter 4 Worksheet D

The cyclohexane derivative shown below in a top view can exist in two interconverting chair forms. Draw each chair (4 pts), and calculate $\Delta G°$ for their interconversion, given that $\Delta G°_{EQ \rightarrow AX}(OH) = 2.2$ kJ/mol, $\Delta G°_{EQ \rightarrow AX}(CH(CH_3)_2) = 8.8$ kJ/mol, and $\Delta G°_{EQ \rightarrow AX}(F) = 1.0$ kJ/mol (2 pts). Circle the more stable conformation (2 pts), and ESTIMATE K_{EQ} (2 pts). (BONUS: If hydrogen bonds between OH and F are worth 12 kJ/mol, what is K_{EQ}? (2 pts))

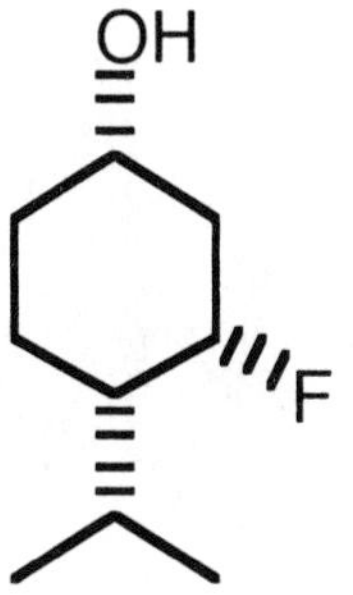

Chapter 4 Worksheet E

a) Consider only those isomers of C_5H_8 that contain a single 3-membered ring. Draw all eight of these isomers (12 pts).

b) C_5H_8 also has several bicyclic isomers, like X and Y below. Identify each distinct hydrogen in X and Y (A --> ?) (6 pts). Draw the one bicyclic isomer of C_5H_8 that has *cis* and *trans* forms (1 pt).

X

Y

Name:_____   Date:_____

Chapter 5 Worksheet A

Sketch all the isomers of $C_4H_{10}O$ (12 pts).

Name:_____   Date:_____

Chapter 5 Worksheet B

Replace each type of hydrogen in the hexane isomer below with a SINGLE chlorine
atom, to derive several of the isomers of $C_6H_{13}Cl$ (8 pts). Draw only one member of each
enantiomeric pair. Circle the chiral isomers (4 pts), and use the R/S nomenclature to label
their chiral centers (3 pts). EXTRA CREDIT-- for either achiral isomer, label each
distinct hydrogen A, B, ... (4 pts -- no partial credit).

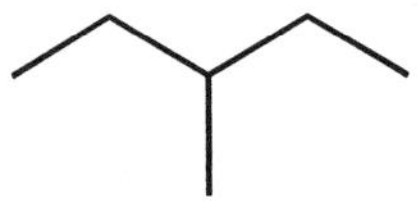

Chapter 5 Worksheet C

One isomer of C_4H_7F is chiral. Six others are a distinct type of stereoisomer. Draw clear 3-D representations of these eight isomers (8 pts), and label each with the appropriate stereochemical descriptor (E/Z or R/S) (4 pts).

Chapter 5 Worksheet D

Three isomers of $C_5H_{11}F$ are chiral. The other five are not, and one of these, **structure X**, has only two distinct types of protons.

a) Draw all EIGHT isomers — draw ONLY ONE MEMBER OF EACH ENANTIOMERIC PAIR, with wedged and dashed bonds at the chiral center (16 pts).

b) Circle the THREE CHIRAL ISOMERS (6 pts), and draw a rectangle around **X** (2 pts).

Chapter 5 Worksheet E

a) Consider only those isomers of C_5H_8 that contain a single 3-membered ring. Two of these isomers are chiral. Draw all eight of these isomers — draw only one member of each enantiomeric pair (16 pts). Circle the two chiral isomers (4 pts).

Chapter 6 Worksheet A

Complete the reactions below by filling in the empty boxes with appropriate structures (15 pts).

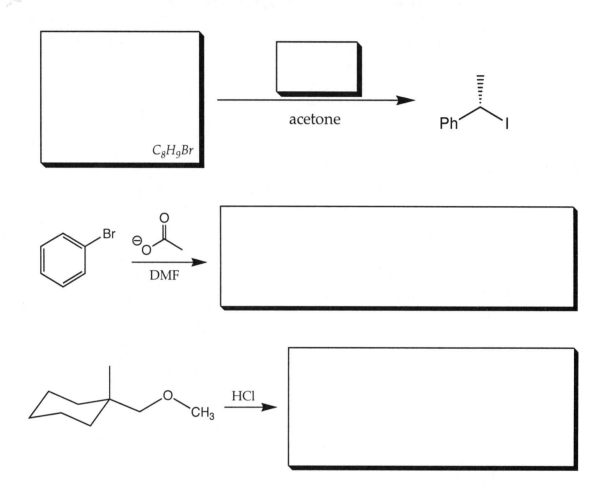

Chapter 6 Worksheet B

An interesting transformation is shown below. Follow the instructions given below to draw a detailed, three-step, curved-arrow mechanism for this reaction, being sure to indicate the stereochemistry of any intermediates (2 pts/step).

a) The OH group is protonated by acid to make it a

_____ (three words, 2 pts).

b) A lone pair on the sulfur atom acts as a nucleophile to make a THREE-MEMBERED ring intermediate.

c) The remaining OH group attacks the TERTIARY carbon to afford the bicyclic product having two FIVE-MEMBERED rings. State in the box what is wrong with the last step as described above (about five words, 2 pts). Indicate the stereochemistry at any chiral centers (4 pts).

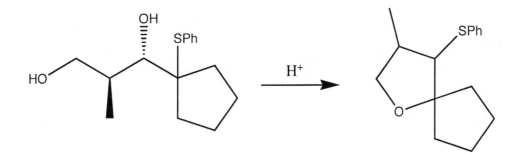

Chapter 6 Worksheet C

The sequence of reactions below affords the two isomeric products shown. Using the correct curved-arrow formalism, and indicating only one step per structure, provide a FOUR-STEP mechanism for the formation of these two products, from a common intermediate **X** (Hint: PROTON TRANSFER REACTIONS ARE VERY FAST!) (8 pts). Why are the two products not formed in equal amounts (2 pts)?

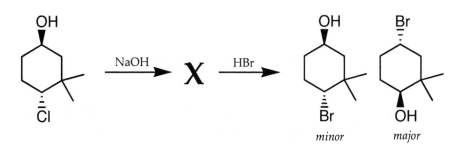

Chapter 6 Worksheet D

Trimethyl phosphite reacts with methyl bromide to afford the product shown, via a TWO-STEP mechanism. Keep in mind that P is below N in the periodic table, and draw a DETAILED, ARROW-PUSHING MECHANISM for this reaction.

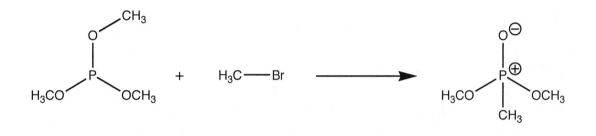

Chapter 6 Worksheet E

The reaction below is 1st order in substrate and 0th order in hydroxide. Propose a DETAILED, ARROW-PUSHING MECHANISM consistent with these facts, and indicate the stereochemistry at any chiral centers (10 pts). Explain why only this structural isomer is obtained (4 pts).

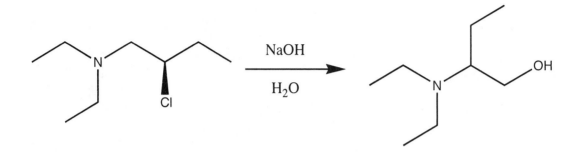

Name:_____   Date:_____

Chapter 7 Worksheet A

a) Correct any errant abbreviations for reaction mechanisms below (5 pts). Match each MECHANISM to its appropriate RXN COORDINATE (10 pts). More than one mechanism may correspond to a single coordinate diagram. Also, pick one word out of the box below that characterizes each mechanism, and place it in the appropriate empty box on the right. Not all the words have to be used; a word may be used more than once (5 pts).

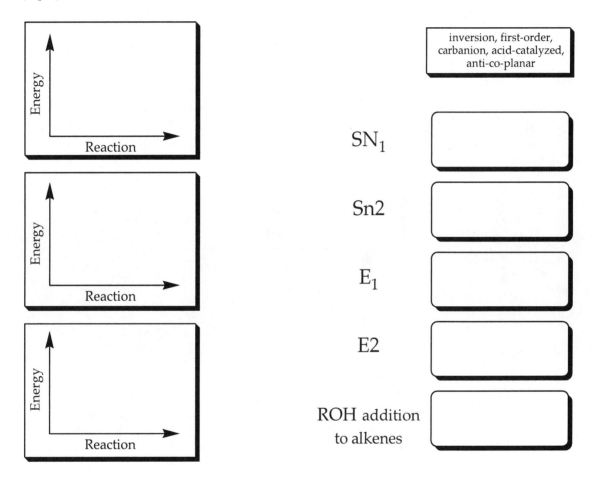

inversion, first-order, carbanion, acid-catalyzed, anti-co-planar

SN_1

$Sn2$

E_1

$E2$

ROH addition to alkenes

b) The decision between S_N2 and E2 is often a difficult one. Put a check into the appropriate box, to indicate whether the conditions on the left favor S_N2 or E2 chemistry (5 pts).

CONDITIONS	S_N2	E2
Bulky nucleophile/base		
I^- as nucleophile		
3° carbon w/leaving group		
1° carbon w/leaving group		
Small nucleophile/base		

Chapter 7 Worksheet B

Secondary alkyl halides undergo both S_N1 and S_N2 substitution, depending on the nature of the solvent and nucleophile. Provide DETAILED, ARROW-PUSHING MECHANISMS for the two reactions below (6 pts). Draw each product, including stereochemistry at any chiral centers (3 pts).

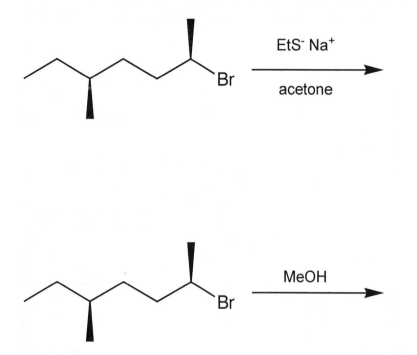

Name:_____ Date:_____

Chapter 7 Worksheet C

Choose letters from the box below that describe each of the five isomeric bromides below. More than one letter will apply to each compound.

A = one chiral center
B = two chiral centers
C = 1°
D = 3°
E = β-branched
F = has enantiomers, but no diastereomers
G = has diastereomers
H = slowest S_N2 + no E2
I = fast S_N2
J = fast S_N1
K = S_N1 with rearrangement

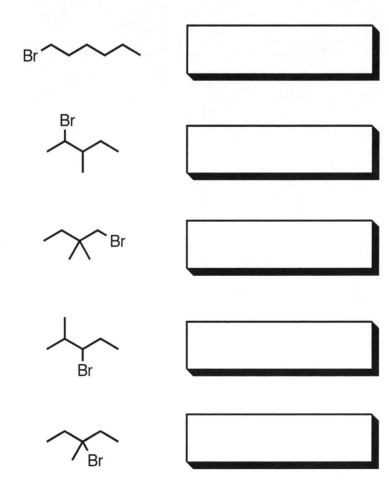
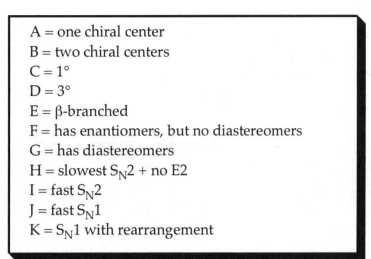

Chapter 7 Worksheet D

On the template provided, draw a Newman projection, looking down C3-C4, of **E** (4 pts). Draw the most reactive conformer of **E**, under E2 conditions (3 pts). Draw the product of the E2 reaction (3 pts).

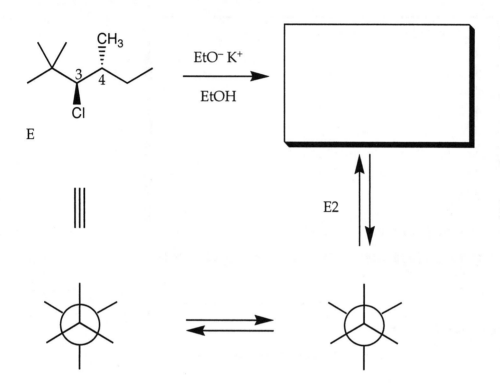

Chapter 7 Worksheet E

a) Draw the cyclohexane below in its lowest energy CHAIR conformation (4 pts).

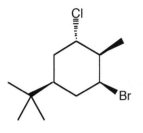

b) On treatment with a strong, hindered base, a single product is obtained. Use the chair form above to provide a DET.\!LED, ARROW-PUSHING MECHANISM for formation of this product (4 pts).

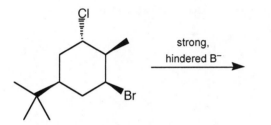

Chapter 8 Worksheet A

Treatment of the first compound below with aqueous acetone affords a solution with no optical rotation, while the same conditions afford a chiral product in the case of the second compound. Explain by providing a mechanism for each reaction.

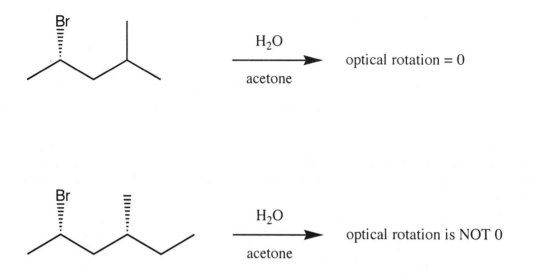

H_2O / acetone → optical rotation = 0

H_2O / acetone → optical rotation is NOT 0

Chapter 8 Worksheet B

Provide a DETAILED, ARROW-PUSHING MECHANISM for the transformations below (10 pts).

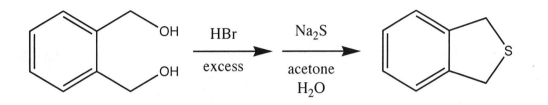

Name:_____   Date:_____

Chapter 8 Worksheet C

Propose a synthesis of the ether shown below from the starting materials provided and any inorganic reagents, acids, or bases.

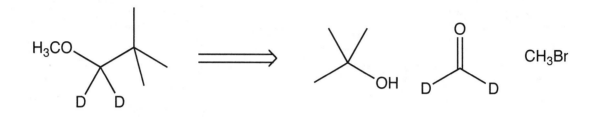

Name:_____ Date:_____

Chapter 8 Worksheet D

Synthesize the target below via the synthetic route suggested by the arrows and boxes.
Indicate necessary reactants/intermediates/starting materials in the boxes (8 pts)

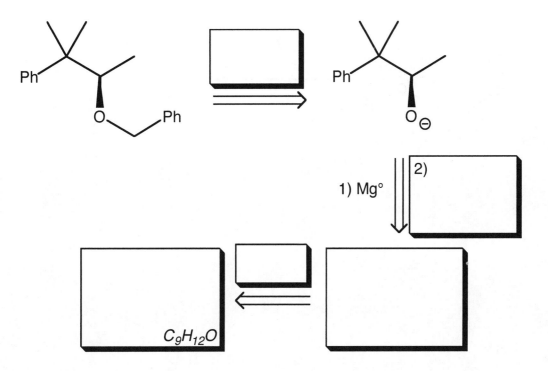

Chapter 8 Worksheet E

Prepare 2-methylpropanal bearing ^{13}C labels at C-1 and C-2 from 2-^{13}C-acetone and ^{13}C-formaldehyde, and any inorganic reagents.

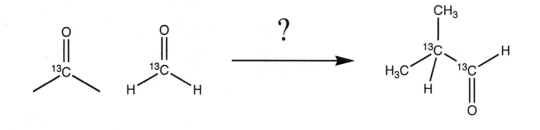

Chapter 9 Worksheet A

Complete the reactions below by filling in the empty boxes with appropriate structures.

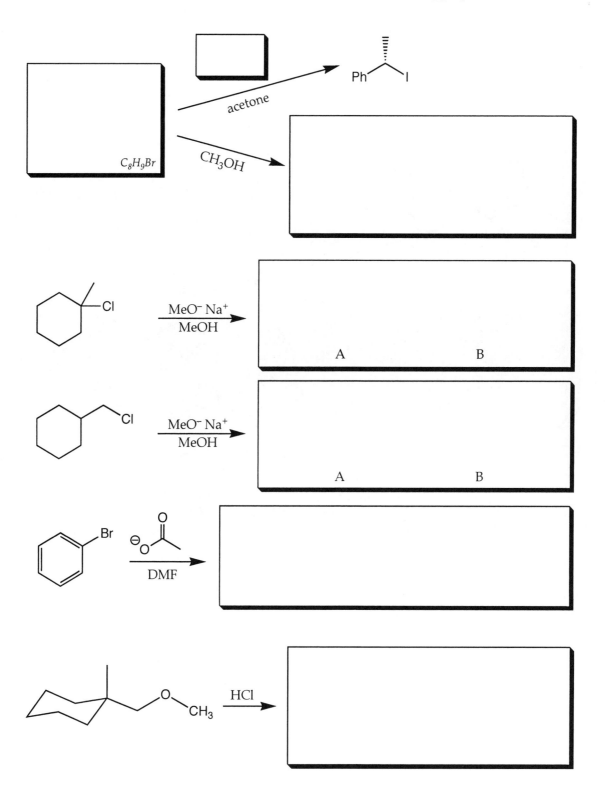

Chapter 9 Worksheet B

Either enantiomer of the THIOETHER below may be prepared from (*S*)-2-butanol. Propose a synthesis of each enantiomer (12 pts).

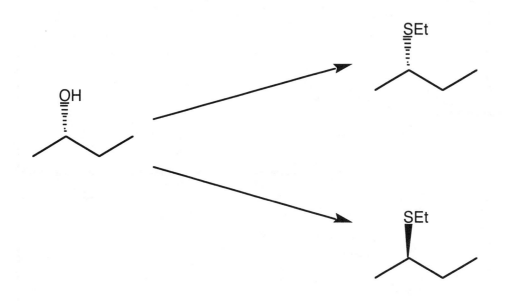

Name:_____   Date:_____

Chapter 9 Worksheet C

a) Isomers **A**, **B**, and **C** react distinctly with KOH in methanol. To understand their behavior, start by drawing the lowest energy chair forms of each isomer in the box below each top view below (9 pts).

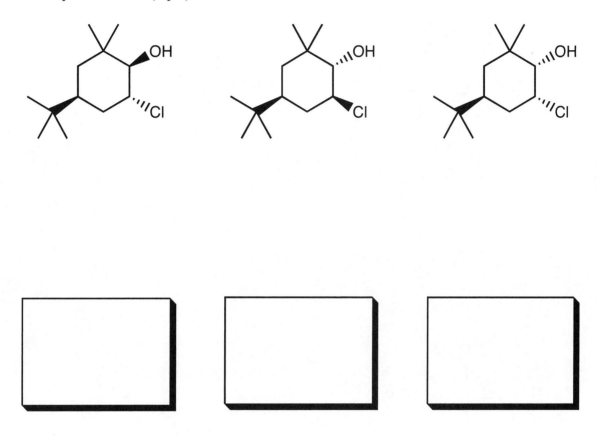

A ===> no epoxide on treatment with KOH/methanol
B ===> epoxide forms rapidly on treatment with KOH/methanol
C ===> epoxide forms slowly on treatment with KOH/methanol

Based on these results, label the boxes above as **A**, **B**, or **C** (3 pts). Explain why **A** does not react (2 pts). Draw the reactive conformation of **C** to explain why it reacts so slowly (3 pts).

Chapter 9 Worksheet D

Using the correct curved-arrow formalism, and indicating only one step per structure, provide the best mechanism for the reaction shown below (3 pts per step).

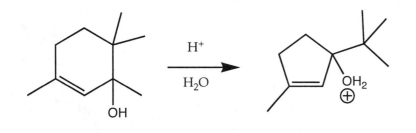

Name:_____   Date:_____

Chapter 9 Worksheet E

Propose a synthesis of the ether shown below from the starting materials provided and any inorganic reagents, acids, or bases.

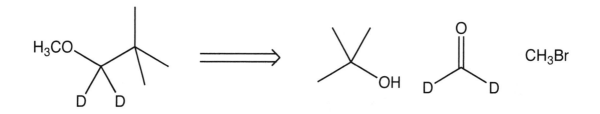

Name:_____   Date:_____

Chapter 10 Worksheet A

The spectrum of a PRIMARY alkyl bromide of formula $C_5H_{11}Br$ is shown, as acquired by a 60 MHz NMR spectrometer. What TWO CONNECTIVITIES can you infer from this low-resolution spectrum (4 pts)? Subtract these pieces from the formula given to complete the structure of this isomer (4 pts).

What would the overlapping resonances near 1.75 ppm look like when examined at higher resolution with a 400 MHz NMR spectrometer (4 pts)? List them from high to low ppm (e.g., (d, 1H), (t, 3H)).

Chapter 10 Worksheet B

The ^{1}H NMR spectrum of cyclohexene includes a 2H triplet at 5.6 ppm. The spectrum of dihydropyran, **X**, includes a doublet at 6.4 and a doublet of triplets at 4.7 ppm.

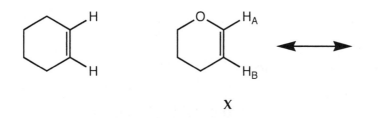

X

Connect H_A and H_B to the appropriate chemical shift value, based on the coupling patterns given (4 pts).

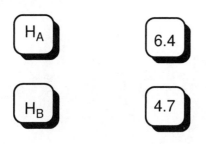

Draw a resonance form for **X**, next to its structure above (3 pts).
Use this structure to explain the chemical shift of HB (3 pts).

Chapter 10 Worksheet C

Infer the connectivity/key functional group of the product of the reaction below from its
¹H NMR data ((d,1H) ==> **CH-CH**, etc., 4 pts). Provide a DETAILED, ARROW-
PUSHING MECHANISM for this reaction (6 pts).

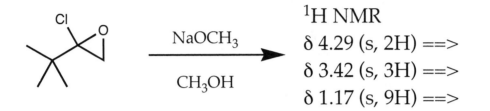

NaOCH₃

CH₃OH

¹H NMR

δ 4.29 (s, 2H) ==>

δ 3.42 (s, 3H) ==>

δ 1.17 (s, 9H) ==>

Name:_____   Date:_____

Chapter 10 Worksheet D

Infer the connectivity/key functional group of the product of the reaction below from its
^{1}H NMR data ((d, 2H) ==> **CH2**-CH, etc., 9 pts). Provide a DETAILED, ARROW-
PUSHING MECHANISM for this reaction (6 pts).

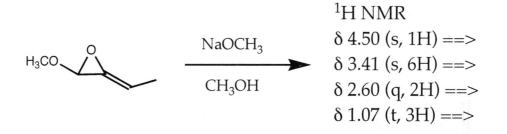

^{1}H NMR

δ 4.50 (s, 1H) ==>

δ 3.41 (s, 6H) ==>

δ 2.60 (q, 2H) ==>

δ 1.07 (t, 3H) ==>

Chapter 10 Worksheet E

Amide **Y** has TWO SET OF PEAKS CORRESPONDING TO ETHYL GROUPS in its
¹H NMR spectrum, AND TWO PEAKS CORRESPONDING TO METHYL GROUPS,
instead of just one of each. At high temperature (>100 °C), one observes only one set of
each.

Explain, with a clear 3-D representation of **Y** (4 pts), and any relevant Lewis structures
(4 pts).

Chapter 11 Worksheet A

a) Compound **A**, $C_7H_6O_2$, has an IR band between 3200-3500 cm^{-1}, even at high dilution. The ^{1}H NMR spectrum has a ^{1}H singlet at 9.10 ppm, a doublet, two doublets of doublets, and another doublet, each of ^{1}H, in the aromatic region. There is also a broad 1H resonance in the ^{1}H NMR spectrum. Systematically deduce the substructures present in **A** (8 pts).

IR 3200-3500 cm^{-1} ==>

9.10 ppm ==>

aromatic d ==>

aromatic dd ==>

aromatic dd ==>

aromatic d ==>

broad peak ==>

b) Propose a structure for compound **A** (5 pts).

c) Based on the IR stretch mentioned above, would you expect v_{CO} of **A** to appear at a frequency lower or higher than 1710 cm^{-1} (circle your answer)? Draw a resonance form to help explain why (4 pts). Would you expect the broad resonance in the ^{1}H NMR to be unusually-shielded, or -deshielded (2 pts)?

$v_{CO} < 1710 \ cm^{-1}$

$v_{CO} > 1710 \ cm^{-1}$

broad resonance is

unusually-shielded

unusually-DEshielded

Chapter 11 Worksheet B

For the problems below, please recall that the stretching frequency of a bond is related to its bond order.

a) The carbonyl derivatives shown below have the indicated relative stretching frequencies. Draw a better resonance form for the first structure (4 pts). Explain, in just a few words, the origin of the relative v_{CO}-s (4 pts).

b) The two related ketones shown have distinct IR bands, in the C=O region. Explain, in just a few words and with one drawing, the origin of these stretching frequencies (8 pts).

CH₃   =O   1740 cm⁻¹   CH₃   =O   1700 cm⁻¹

O-CH₃   O-H

c) The two amides below have distinct carbonyl stretching frequencies, with the bicyclic structure having v_{CO} much greater than v_{CO} for the non-cyclic analog. Explain the origin of this effect, with just a pretty picture (4 pts).

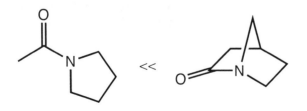

Chapter 11 Worksheet C

The carbonyl stretching frequency of KETENE, $H_2C=C=O$ (drawn below), is 2150 cm^{-1}.
By comparing this to the CO stretching frequency of a KETONE, you would conclude

that the bond is (greater than or less than) _____
a double bond (2 words, 2 pts).

Draw an appropriate minor resonance isomer to justify your answer (2 pts).

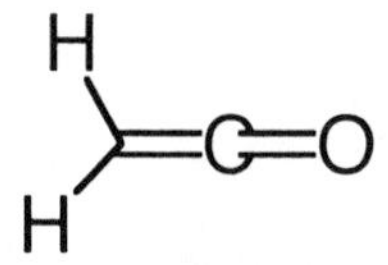

KETENE

In Organic Chemistry II, you'll learn about the reaction of ethanol with ketene, which
affords a compound with an IR band at 1735 cm^{-1}, and ^{1}H NMR resonances at about 4.1
ppm (q, 2H), 2.1 ppm (s, 3H) and 1.3 ppm (t, 3H). What substructure corresponds to each
of the IR and NMR patterns (2 pts each).

1735 cm^{-1} →

(q, 2H) →

(s, 3H) →

(t, 3H) →

Name:_____   Date:_____

Chapter 11 Worksheet D

Protonation of the alkene below (methyl vinyl ether) can give two carbocation products. Draw each product (4 pts each). Circle the more stable cation (3 pts). In the plane provided, draw the orbitals involved in stabilization of the more stable product (4 pts).

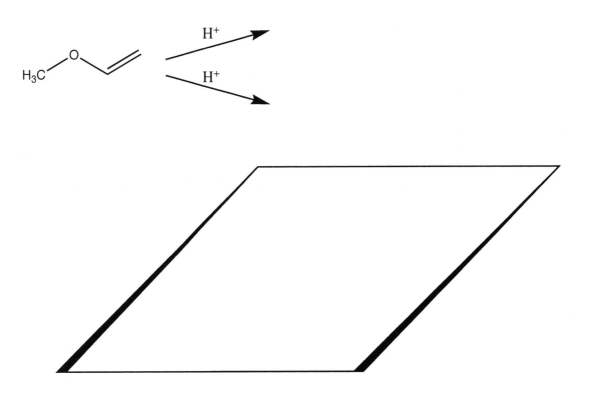

Chapter 11 Worksheet E

Protonation of ketene, followed by addition of water to the resulting cation, ultimately affords acetic acid. Draw the DETAILED, THREE-STEP, ARROW-PUSHING MECHANISM for this reaction (9 pts). Use a relevant resonance form to explain why only this product is obtained (4 pts).

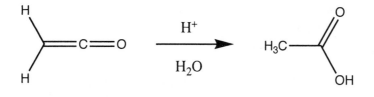

Chapter 12 Worksheet A

Addition of HBr to (E)-3,4-dimethyl-2-butene affords a product having the SEVEN ^{1}H NMR references listed below.

a) Provide a DETAILED, ARROW-PUSHING MECHANISM for this reaction (6 pts).

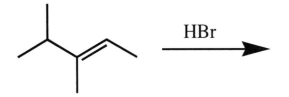

Product ^{1}H NMR spectrum: (septet, 1H), (s, 3H), (dq, 1H), (dq, 1H), (d, 3H), (d, 3H), (t, 3H)

b) Use the diastereotopicity test to rationalize the spectral data (6 pts).

Chapter 12 Worksheet B

Provide a DETAILED, ARROW-PUSHING MECHANISM for the reaction shown below (6 pts). By labeling a key carbon of the first INTERMEDIATE, explain the selectivity of the reaction (1 pt).

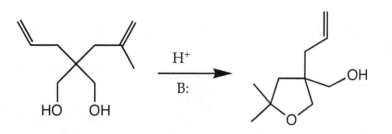

Name:_____   Date:_____

Chapter 12 Worksheet C

Chiral compounds rotate plane polarized light. The acid-catalyzed reaction shown below can be followed by measuring the decrease in optical rotation over time.

i) Add H$^+$ to the alkene to form the most stable cation (3 pts).
ii) Add ethanol to the resulting cation, to form the TWO PRODUCTS of the reaction (3 pts).
iii) Draw proper chair representations of the two products (6 pts), and circle the more stable product (2 pts).
iv) IN THE BOX BELOW, state the property of these products that allows the reaction to be followed in the manner mentioned above (2 pts).

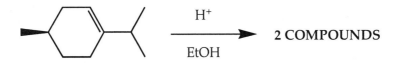
$$\text{H}^+$$
$$\xrightarrow{\hspace{2cm}}$$
EtOH
2 COMPOUNDS

Chapter 12 Worksheet D

a) Protonate the diene below to give the most stable cation (3 pts).

   H-Br ⟶   *one pair of mirror image isomers, and one pair of geometric isomers*

b) Draw any additional resonance forms of this cation (2 pts).

c) Use appropriate arrows to derive all four products of the reaction (10 pts).

d) List the ^{1}H NMR data that you expect for three of the products above, from HIGHEST to LOWEST ppm, including the integration and multiplicity of each peak (e.g., (q, 2H)) (10 pts). Identify any "LARGE" coupling constants by using a "**D**."

Chapter 12 Worksheet E

Addition of DBr to the alkene shown, in the presence of a RADICAL INITIATOR Q•, will afford a mixture of stereoisomeric products. Use proper ARROWS to initiate the addition reaction by generating Br• from DBr (3 pts). Add this radical to the LEAST HINDERED FACE of the alkene (4 pts). Finally, use an appropriate source of D• to complete the reaction, to give a mixture of stereoisomers (4 pts).

Hydrogen had two nuclear spin states ($\pm 1/2$), resulting in the coupling patterns that are normally observed for organic compounds. Deuterium has three possible spin states, which results in different coupling patterns. These can be derived by the same "SPLITTING DIAGRAMS" that are used for routine ^{1}H NMR spectra. Choose either stereoisomer, and given that $J = 2.5$ Hz for coupling of H and D, sketch the pattern that would be observed for the HYDROGEN at C-2 in the PRODUCT. Assume that $J_{AX-AX} = 10$ Hz and that $J_{AX-EQ} = 5$ Hz (6 pts).

Chapter 13 Worksheet A

Fill in the boxes to complete the synthetic sequence below (12 pts).

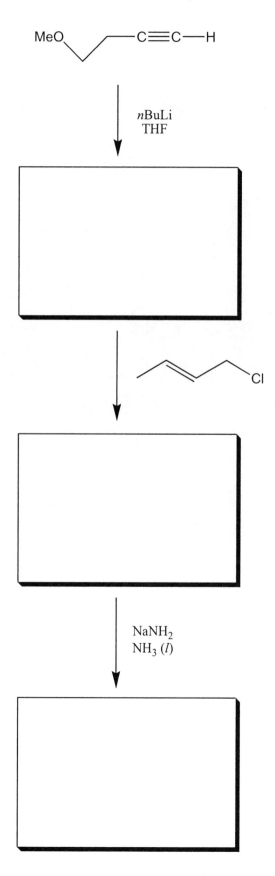

Chapter 13 Worksheet B

With 1-PROPYNE as your ONLY SOURCE OF CARBON, synthesize the alcohol
product below (16 pts).

Name:_____   Date:_____

Chapter 13 Worksheet C

Fill in the intermediates in the synthesis below and provide a DETAILED, ARROW-PUSHING MECHANISM for the formation of the cyclic byproduct (20 pts).

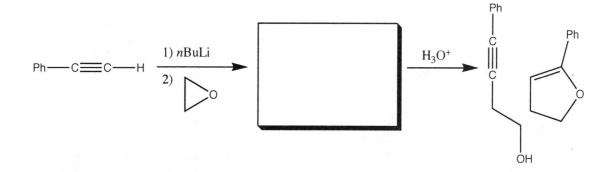

Chapter 13 Worksheet D

Synthesize the ether shown below from the starting materials provided and any inorganic reagents (16 pts).

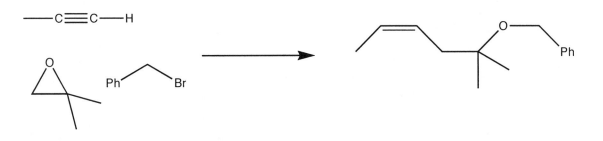

d) List the ^{1}H NMR data that you expect for three of the products above, from HIGHEST to LOWEST ppm, including the integration and multiplicity of each peak (e.g., (q, 2H)) (10 pts). Identify any "LARGE" coupling constants by using a "**D**."

Chapter 13 Worksheet E

Prepare the racemic alcohol below from phenylacetylene, ethyl magnesium bromide, and any inorganic reagents (12 pts).

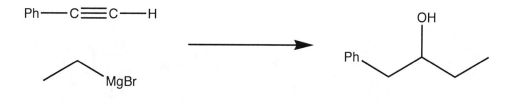

Chapter 14 Worksheet A

Draw a CLEAR, 3-D REPRESENTATION OF THE TRANSITION STATE LEADING
TO INTERMEDIATE X (4 pts), AND THE STRUCTURE OF X (4 pts). Complete the
mechanism by providing a DETAILED, ARROW-PUSHING MECHANISM for the
transformation below (8 pts).

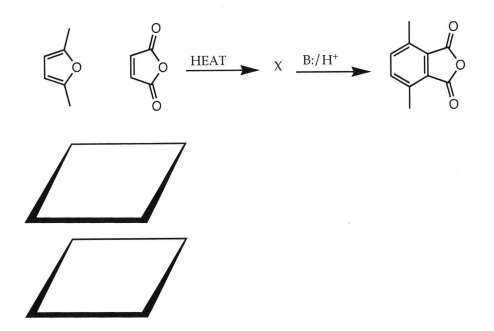

Chapter 14 Worksheet B

RETROSYNTHESIZE the target below in about 5-6 steps, from the starting material
suggested by the formula below, and any additional compound having fewer than EIGHT
CARBONS (15 pts). Indicate any stereochemistry of all structures throughout the
synthesis (6 pts). Draw a box around any intermediate(s) that have THREE 3H triplets in
the ^{1}H NMR (4 pts).

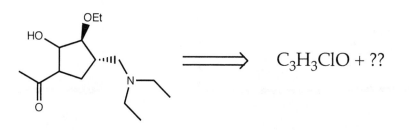

C_3H_3ClO + ??

Name:_____   Date:_____

Chapter 14 Worksheet C

In the small box, identify the number of π electrons involved in the electrocyclic reaction below, and state whether it proceeds via a CON or DISrotatory process (4 pts). Draw the product of the reaction, as well as the major stereoisomer obtained on epoxidation of this compound (6 pts).

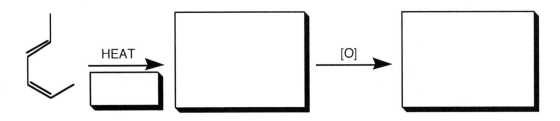

Chapter 14 Worksheet D

1,3-cyclooctadiene is deprotonated at one site, affording an anion that rearranges
THERMALLY to afford a specific stereoisomeric product. How many π electrons are
involved in this step (4 pts)? Describe the mechanism of the process with appropriate
terms (4 pts). This anion can be trapped by CO_2 to give the acid shown. Draw the relative
stereochemistry of the final product (2 pts).

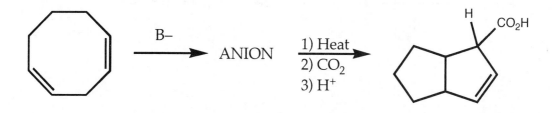

Name:_____   Date:_____

Chapter 14 Worksheet E

Derive the structures of the key intermediate below, and the structure of the final product (6 pts). How many π electrons are involved in the final step (2 pts)? Write an appropriate word IN THE BOX to describe the mechanism of this step (2 pts).

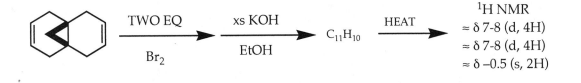

TWO EQ → xs KOH → $C_{11}H_{10}$ → HEAT →

Br$_2$ EtOH

^{1}H NMR
≈ δ 7-8 (d, 4H)
≈ δ 7-8 (d, 4H)
≈ δ –0.5 (s, 2H)

Name:_____   Date:_____

Chapter 15 Worksheet A

Circle the aromatic compounds from among the structures below (12 pts).

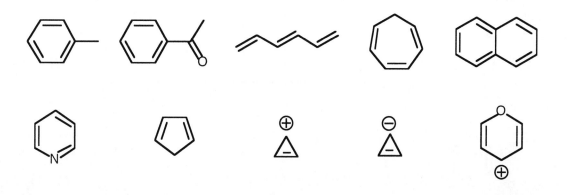

Chapter 15 Worksheet B

γ-pyrone is "unusually-stable" and has a remarkable dipole moment compared to cyclohexanone. In the plane provided, draw a 3-D representation of γ-pyrone, showing relevant π-bonds and lone pairs (6 pts). Draw a top view of an important resonance form of γ-pyrone (4 pts).

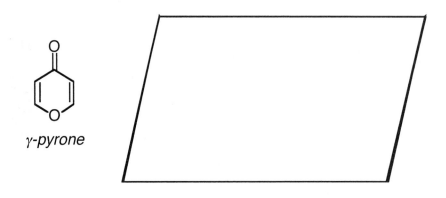

γ-pyrone

Name:_____   Date:_____

Chapter 15 Worksheet C

The bicycle below can be deprotonated TWICE, to afford an unusually stable product.
Draw the product of the first deprotonation (3 pts). Draw the BEST RESONANCE
FORM of the product of the second deprotonation (3 pts).

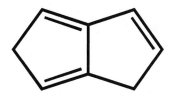

Chapter 15 Worksheet D

Methyl vinyl ketone readily incorporates deuterium on treatment with base in D_2O, but cyclobutenone does not. Draw a chemical structure to explain this observation (8 pts).

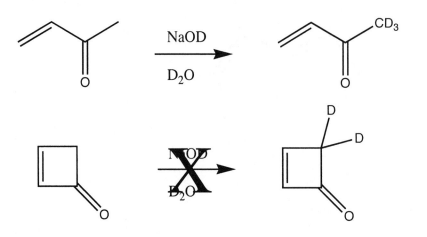

Chapter 15 Worksheet E

Treatment of the triene below with acid while heating leads to the appearance of a single methyl resonance in the ^1H NMR spectrum. How many methyl peaks would you expect for the product of this reaction (4 pts)? Explain with a MECHANISM why only one peak is observed (8 pts).

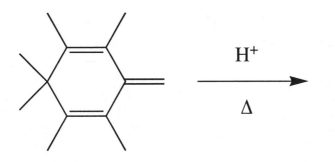

Chapter 16 Worksheet A

Draw the product of addition of ONE EQUIVALENT of ^{t}butyl chloride to EACH of the two aromatic rings of the substrate below (10 pts). Draw resonance forms of any possible intermediates to explain the selectivity of addition to each ring (5 pts each).

2 EQUIVALENTS
^{t}butyl chloride

AlCl$_3$

Chapter 16 Worksheet B

Ibuprofen can be prepared via two routes that involve the bromide below as a key intermediate. Synthesize this bromide, starting from styrene and any compounds having four or fewer carbons (6 pts). Draw a resonance form of a key intermediate to explain the regiochemistry of the E^+ addition reaction (4 pts).

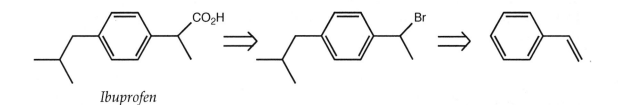

Ibuprofen

Name:_____   Date:_____

Chapter 16 Worksheet C

Anthracene reacts first with O_2N^+ and then with Cl^- to give a compound with the formula of anthracene + NO_2 + Cl ($C_{14}NO_2Cl$). Treatment of this product with NaOH affords $C_{14}H_9NO_2$.

Draw the structure of the most stable product possible from O_2N^+ addition (5 pts). Why is this cation the favored addition product (5 pts)? Provide a DETAILED, ARROW-PUSHING MECHANISM for the reaction of the intermediate with NaOH, and draw the structure of the final product (5 pts).

$$\xrightarrow[Cl^-]{O_2N^+} \quad C_{14}H_{10}NO_2Cl \quad \xrightarrow{NaOH} \quad C_{14}H_9NO_2$$

anthracene

Chapter 16 Worksheet D

Provide a DETAILED, ARROW-PUSHING MECHANISM FOR the "NIH SHIFT" (10 pts). Draw an important resonance form of a key intermediate to justify the regiochemistry of this reaction (3 pts).

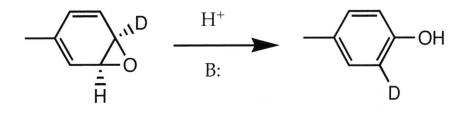

Chapter 16 Worksheet E

Derive the substructures of a KEY INTERMEDIATE in the synthesis of the target below from the ^{1}H NMR data provided (7 pts). RETROSYNTHESIZE the target molecule from methoxybenzene and any other compounds having SIX carbons or fewer (9 pts).

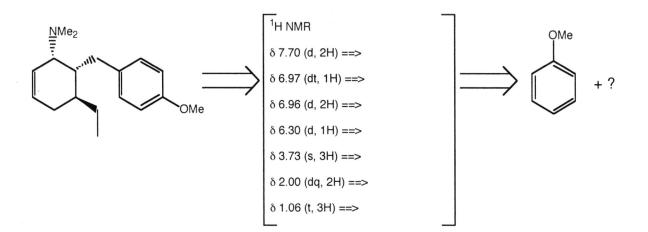

^{1}H NMR

δ 7.70 (d, 2H) ==>

δ 6.97 (dt, 1H) ==>

δ 6.96 (d, 2H) ==>

δ 6.30 (d, 1H) ==>

δ 3.73 (s, 3H) ==>

δ 2.00 (dq, 2H) ==>

δ 1.06 (t, 3H) ==>

Name:_____   Date:_____

Chapter 17 Worksheet A

Complete the synthesis below by providing appropriate reagents/intermediates in the boxes (8 pts). Complete the structure of the product of the synthesis, including stereochemistry (4 pts).

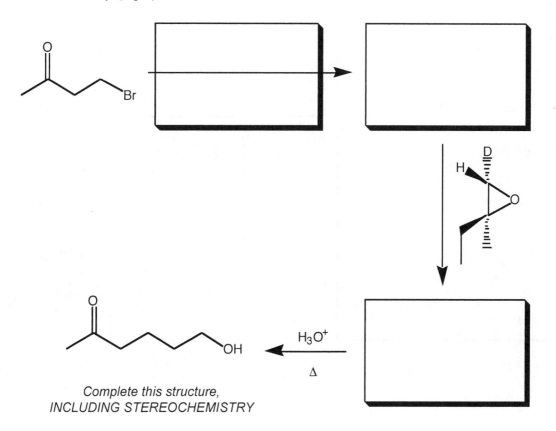

Complete this structure,
INCLUDING STEREOCHEMISTRY

Chapter 17 Worksheet B

Infer the connectivity of the molecule in the box from the C-H ^{1}H NMR resonances listed below (e.g., (t, 3H) ➔ **H₃C-CH₂**) (6 pts). Deduce the structure of the unknown molecule (4 pts).

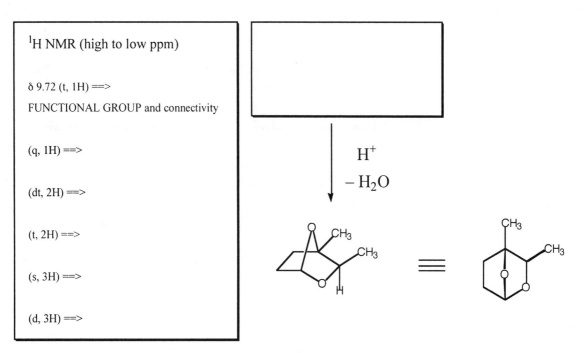

^{1}H NMR (high to low ppm)

δ 9.72 (t, 1H) ==>

FUNCTIONAL GROUP and connectivity

(q, 1H) ==>

(dt, 2H) ==>

(t, 2H) ==>

(s, 3H) ==>

(d, 3H) ==>

H⁺

− H₂O

Chapter 17 Worksheet C

Provide a DETAILED, ARROW-PUSHING MECHANISM for the transformation
shown below (16 pts).

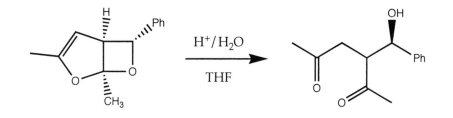

Chapter 17 Worksheet D

Phil Baran's lab at the Scripps Research Institute has prepared *Sceptrin*, a potent anti-viral/antibiotic agent, via the reaction in the box in a key step (*J. Am. Chem. Soc.* **2004**, *126*, 3726-7). Provide a DETAILED, ARROW-PUSHING MECHANISM for this process (10 pts, just write IMINE for any relevant steps). Draw the best resonance form of the product and explain why it is so stable (6 pts).

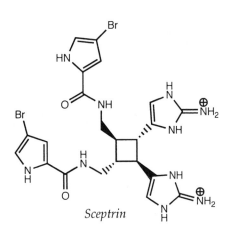

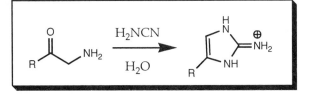

Sceptrin

Name:_____   Date:_____

Chapter 17 Worksheet E

The product below is an intermediate in a synthesis of penicillin. Provide a DETAILED, ARROW-PUSHING MECHANISM for this reaction (12 pts).

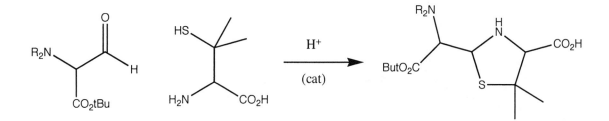

Chapter 18 Worksheet A

Draw the product of treatment of the enone below with basic D_2O (7 pts). Provide a
DETAILED, ARROW-PUSHING MECHANISM for INCORPORATION OF **ANY**
ONE DEUTERIUM (6 pts).

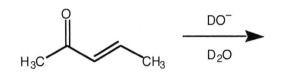

Chapter 18 Worksheet B

The ketone below incorporates FOUR DEUTERIUM ATOMS on treatment with alkoxide in DEUTERATED ethanol. DRAW THE PRODUCT OF H/D EXCHANGE (4 pts), and provide a DETAILED, ARROW-PUSHING MECHANISM for exchange of EACH TYPE OF HYDROGEN (10 pts).

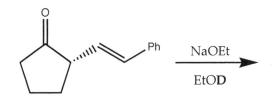

Chapter 18 Worksheet C

Provide a DETAILED, ARROW-PUSHING MECHANISM for the reaction below
(13 pts).

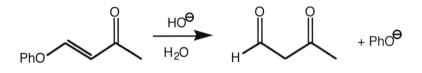

Name:_____   Date:_____

Chapter 18 Worksheet D

Starting materials for the synthesis of the product shown may be inferred from the ^{1}H
NMR data provided. Draw Cmpd A in the box on the right (4 pts). Deduce the structure
of Cmpd B by inferring the detailed connectivity responsible for each of the peaks
described below (4 pts). Identify the functional group responsible for the most deshielded
resonance (2 pts). Justify the relative chemical shift of the next-most-deshielded proton
(2 pts).

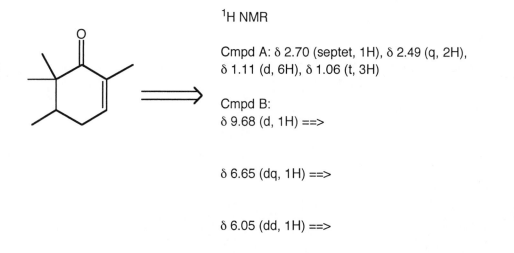

^{1}H NMR

Cmpd A: δ 2.70 (septet, 1H), δ 2.49 (q, 2H),
δ 1.11 (d, 6H), δ 1.06 (t, 3H)

Cmpd B:
δ 9.68 (d, 1H) ==>

δ 6.65 (dq, 1H) ==>

δ 6.05 (dd, 1H) ==>

δ 1.71 (d, 3H) ==>

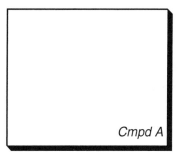

Cmpd A

Chapter 18 Worksheet E

The product below can be synthesized from the cyclopentadiene derivative provided in three to four steps, after reaction with a compound having an IR band at 1705 cm^{-1} and a 3H singlet in its ^{1}H NMR at 3.8 ppm. Indicate conditions required to accomplish each step in the synthesis (10 pts). For an EXTRA 5 pts, label the most deshielded hydrogen, and count the number of distinct hydrogens in the product.

Chapter 19 Worksheet A

Propose a THREE-STEP synthesis of the product below from benzoic acid (12 pts).

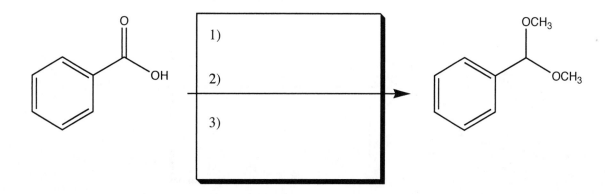

Name:_____   Date:_____

Chapter 19 Worksheet B

Provide reagents to complete the synthesis described below (12 pts).

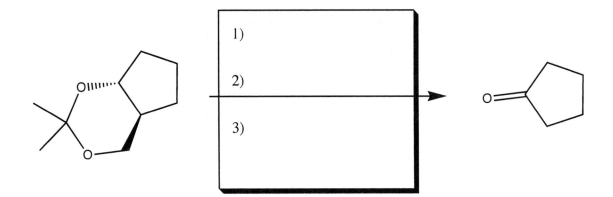

1)

2)

3)

Name:_____ Date:_____

Chapter 19 Worksheet C

Provide a DETAILED, ARROW-PUSHING MECHANISM for the reaction below (10 pts). Use B: for any required bases, and write ENOL over the arrow for any enol/keto isomerizations.

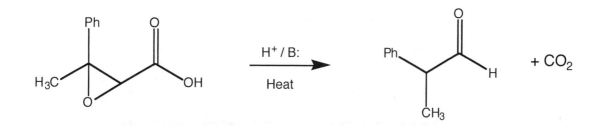

Chapter 19 Worksheet D

Oxidative workup of ozonolysis affords carboxylic acids from substrates like the one shown IN THE BOX. This reaction may be used in a key step in the synthesis of the target molecule shown below. Complete the synthesis by filling in the empty boxes below.

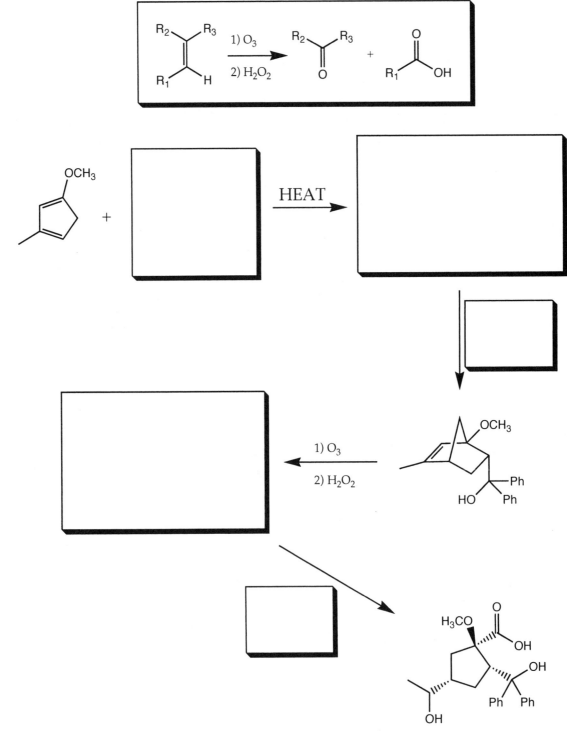

Name:_____ Date:_____

Chapter 19 Worksheet E

Fill in the boxes (5 pts each, 10 pts total). The new reaction in the box below will be helpful. Provide a DETAILED, ARROW-PUSHING mechanism for the THIRD step. Heteroatomic proton transfers may be described at ~H⁺.

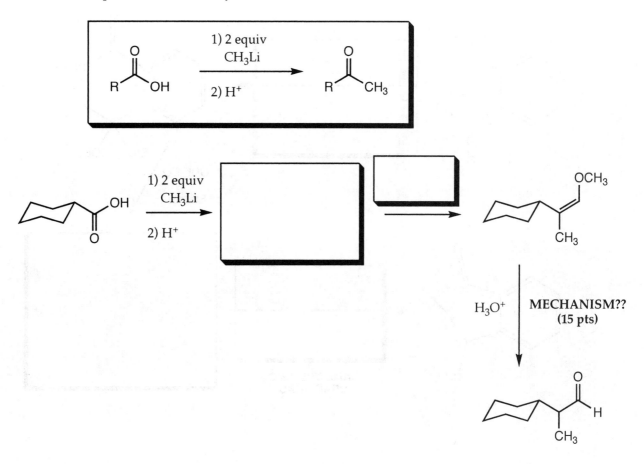

Name:_____   Date:_____

Chapter 20 Worksheet A

Provide reagents for the first transformation (five steps, 8 pts). Draw the intermediate in the empty box (3 pts). NAME the two reactions that afford the final product (4 pts).

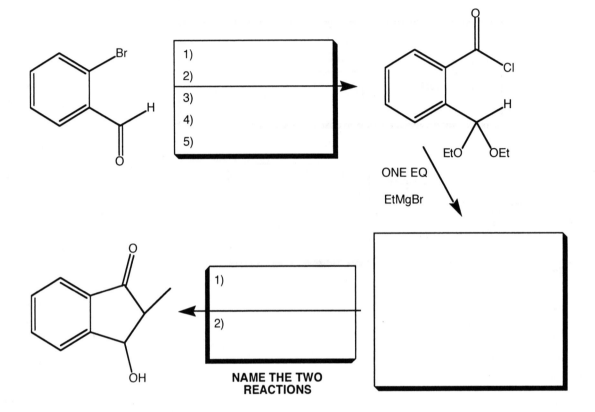

ONE EQ

EtMgBr

NAME THE TWO
REACTIONS

Chapter 20 Worksheet B

Determine the structure of the product from hydrogenation of the enone below, paying close attention to the ^{1}H NMR data provided.

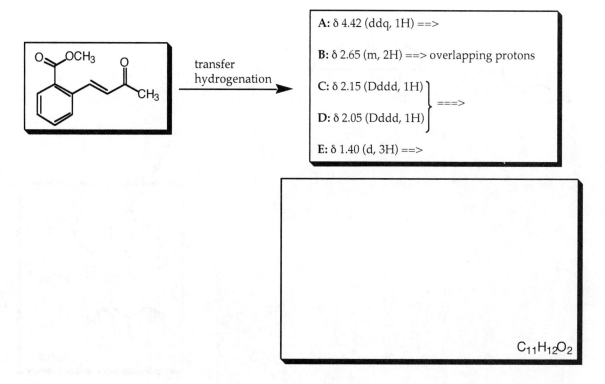

A: δ 4.42 (ddq, 1H) ==>

B: δ 2.65 (m, 2H) ==> overlapping protons

C: δ 2.15 (Dddd, 1H) ⎤
                                  ⎬ ===>

D: δ 2.05 (Dddd, 1H) ⎦

E: δ 1.40 (d, 3H) ==>

$C_{11}H_{12}O_2$

Chapter 20 Worksheet C

The first step in the synthesis of *Calanolide A* is shown below. Provide a DETAILED, ARROW-PUSHING MECHANISM for ONLY THIS STEP.

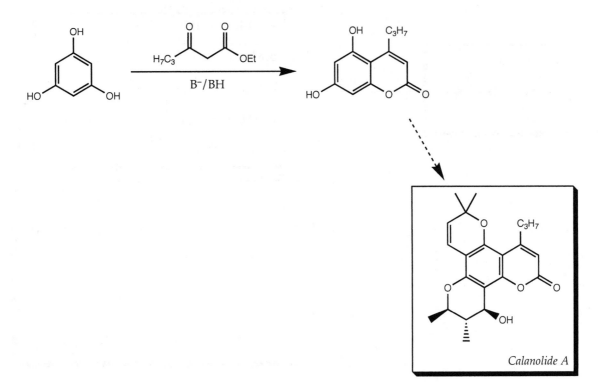

Calanolide A

Chapter 20 Worksheet D

Justin DuBois at Stanford University reported the reaction below (*J. Am. Chem. Soc.*
2003, *125*, 11510-11511*)* as part of an efficient synthesis of *Tetrodotoxin*, the lethal
poison found in the Japanese delicacy *fugu (Blowfish)*. Label all the Electrophilic and
potentially Nucleophilic sites in the STARTING MATERIALS (6 pts). Provide a
DETAILED, ARROW-PUSHING MECHANISM for this transformation (18 pts).

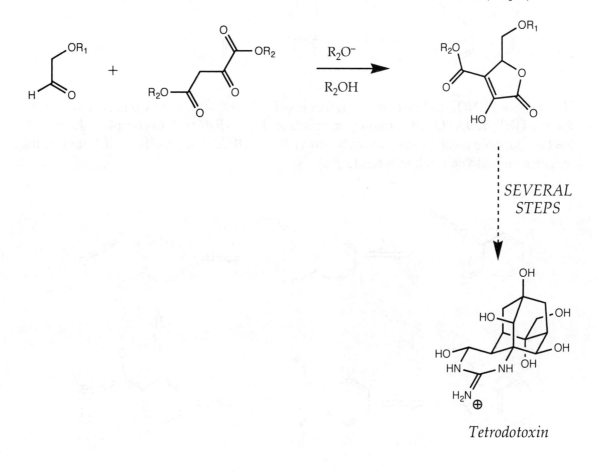

Tetrodotoxin

Chapter 20 Worksheet E

Professor Kresge of U Toronto and co-workers in Switzerland and Hong Kong recently discussed the reaction below (*J. Am. Chem. Soc.* **2003**, *125*, 12872-80).

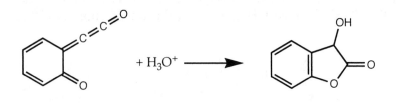

Use the word ENOL to label any steps that involve ENOL-->KETO isomerization, ~H+ for any HETEROATOMIC proton transfers, and draw ARROWS to complete the rest of the two possible mechanisms shown below (17 pts). DRAW A RECTANGLE around the mechanism that you think is better (3 pts).

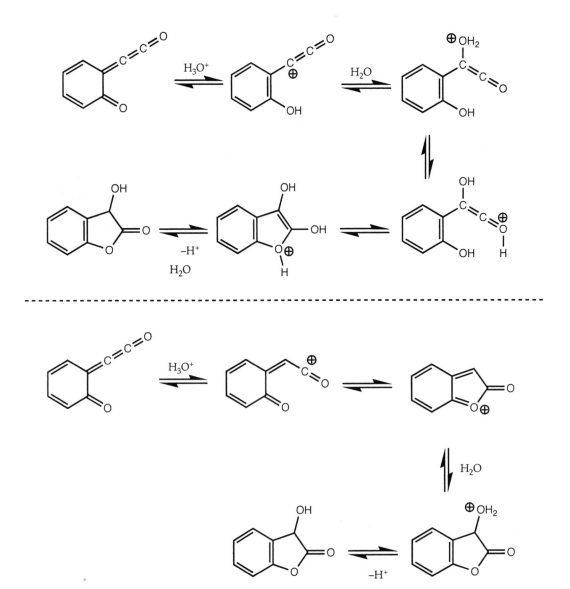

Chapter 21 Worksheet A

USE CHEMICAL STRUCTURES to help explain the relative acidities of the three molecules below (12 pts). In your analyses, emphasize the factors that stabilize/destabilize the product of acid dissociation.

$pK_a = 0.8$ $pK_a = 5.3$ $pK_a = 11.0$

Name:_____   Date:_____

Chapter 21 Worksheet B

Rank the structures below from MOST acidic to LEAST acidic, IN THE BOXES
PROVIDED (1 = most, 4 = least acidic, 6 pts), draw TWO IMPORTANT RESONANCE
FORMS of the CONJUGATE BASE of the most acidic compound (4 pts). Write ONE or
TWO words in the large box to explain the effect of the trifluoromethyl group (2 pts).

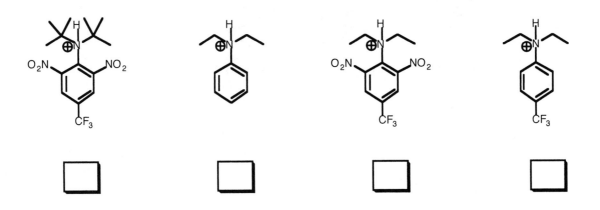

[ ]   [ ]   [ ]   [ ]

[   ]

Chapter 21 Worksheet C

Using the correct curved-arrow formalism, and indicating only one step per structure, provide a TWO-STEP mechanism for the reaction shown below (6 pts). IN THE BOX, suggest why the first step is required for the second step to proceed (4 pts).

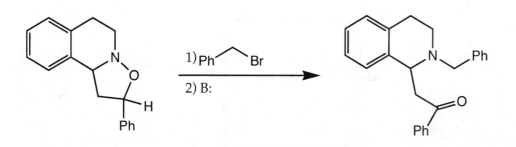

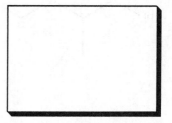

Chapter 21 Worksheet D

Apetinil is the active component of a common diet suppressant. It can be synthesized via several routes. Complete the synthesis of *Apetinil* from the starting materials provided (18 pts). Put a star in the box next to the best synthesis (2 pts).

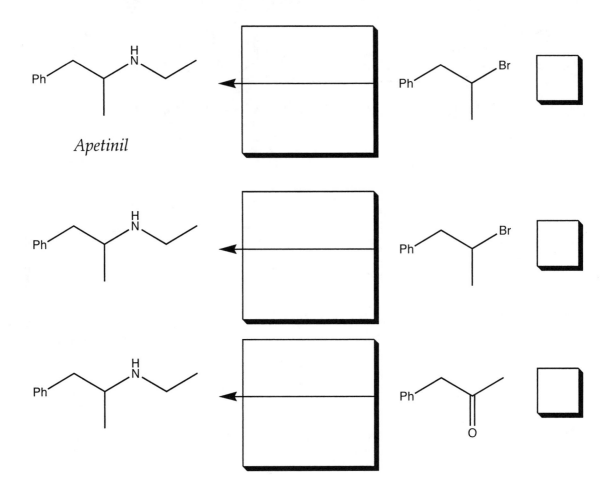

Apetinil

Chapter 21 Worksheet E

Provide a DETAILED, ARROW-PUSHING MECHANISM for the transformation below (16 pts).

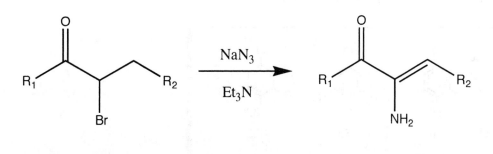

Chapter 22 Worksheet A

a) The following trends are observed, for pKas of 4-substituted PHENOLS. Draw the best resonance form for the anions obtained on deprotonation of 4-cyanophenol and 4-nitrophenol (R=H) to help explain the trend observed (6 pts).

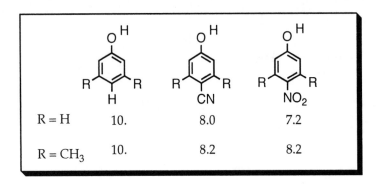

R = H	10.	8.0	7.2
R = CH$_3$	10.	8.2	8.2

Draw a clear 3-D representation of 3,5-dimethyl-4-nitrophenol using the plane provided (4 pts) to help explain why 3,5-dimethyl PHENOLS do not follow this trend.

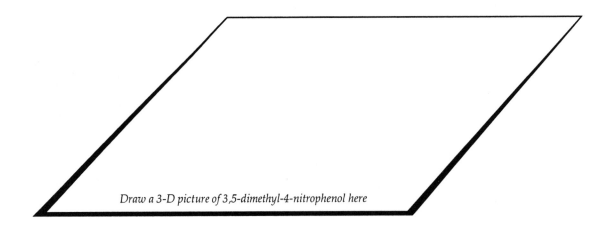

Draw a 3-D picture of 3,5-dimethyl-4-nitrophenol here

Chapter 22 Worksheet B

a) In STRONGLY ACIDIC MEDIA, *N,N*-dimethylaniline affords a bromination product having TWO DOUBLETS, ONE DOUBLET OF DOUBLETS, AND A SINGLET in its ^{1}H NMR spectrum. Provide a DETAILED, ARROW-PUSHING MECHANISM for this reaction (8 pts).

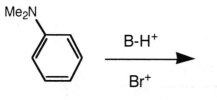

$$\xrightarrow[\text{Br}^+]{\text{B-H}^+}$$

b) The neutralized product from a) can be extracted into ether. The target below can be prepared from the starting materials suggested on the right. RETROSYNTHESIZE this molecule, using an ACID CHLORIDE in a key step (9 pts + 4 pts for analysis of ^{1}H NMR of B).

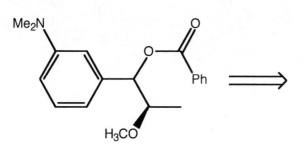

neutralized product of a),
any inorganic reagents,
UNK B, with
^{1}H NMR
δ 9.72 (d, 1H) ==>
δ 3.88 (dq, 1H) ==>
δ 3.24 (s, 3H) ==>
δ 1.32 (d, 3H) ==>
and an acid chloride

Chapter 22 Worksheet C

Nucleophilic aromatic substitution proceeds through attack of a nucleophile on an aromatic ring to afford a resonance-stabilized intermediate, followed by loss of a leaving group to regenerate the aromatic ring. Draw a DETAILED, ARROW-PUSHING MECHANISM FOR the nucleophilic aromatic substitution reaction below (ONLY ONE CHLORINE IS DISPLACED, 6 pts). Draw the best resonance form of the intermediate in the mechanism of this reaction (4 pts).

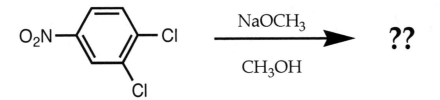

Chapter 22 Worksheet D

Draw the product of Br$^+$ addition and use appropriate resonance forms to deduce the directing properties of each substituent (9 pts).

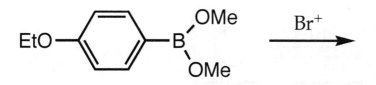

Chapter 22 Worksheet E

Use the information in the box to DEDUCE THE STRUCTURES OF THE
INTERMEDIATES.

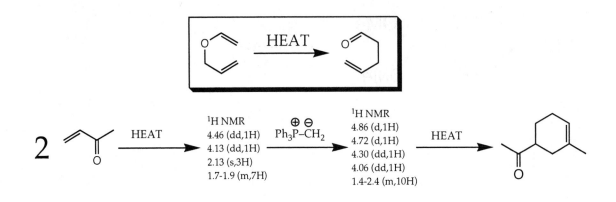

Name:_____   Date:_____

Chapter 23 Worksheet A

Draw the intermediates and final product of the reaction scheme shown below (12 pts).
Indicate the stereochemistry of the major diastereomeric product (4 pts).

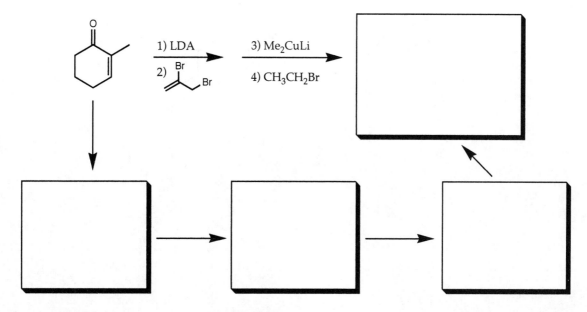

Chapter 23 Worksheet B

Prepare the carboxylic acid PRODUCT from diethylmalonate using appropriate reagents (20 pts).

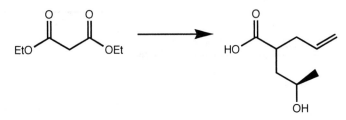

Name:_____   Date:_____

Chapter 23 Worksheet C

Provide a DETAILED, ARROW-PUSHING MECHANISM for the transformation below (12 pts).

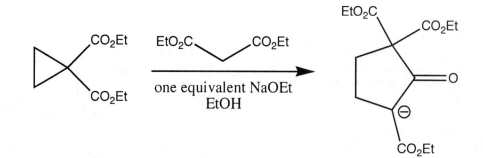

Cmpd A

Name:_____   Date:_____

Chapter 23 Worksheet D

Provide a DETAILED, ARROW-PUSHING MECHANISM for the transformation below (20 pts).

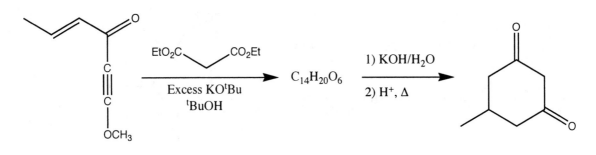

EtO$_2$C   CO$_2$Et

Excess KO^{t}Bu
^{t}BuOH

C$_{14}$H$_{20}$O$_6$

1) KOH/H$_2$O

2) H$^+$, Δ

Chapter 23 Worksheet E

Devise a multi-step synthesis of the amide below from diethyl malonate and any compounds having seven or fewer carbons (16 pts).

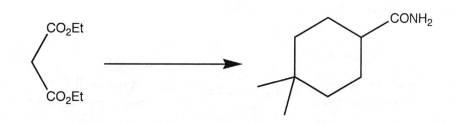

Chapter 24 Worksheet A

$\Delta G°_{EQ \to AX}$ (OH) ≈ 1.0 kcal/mol in cyclohexanol. $\Delta G°_{EQ \to AX}$ (OH) ≈ 0.4 kcal/mol for the C-1 hydroxyl of glucose. Draw the orbitals that are involved in stabilization of α-glucose (4 pts). Provide a DETAILED, ARROW-PUSHING MECHANISM for the base-catalyzed isomerization of α- to β-glucose (6 pts).

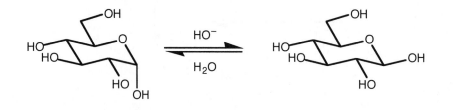

Name:_____   Date:_____

Chapter 24 Worksheet B

Provide a DETAILED, ARROW-PUSHING MECHANISM for the transformation below
(12 pts).

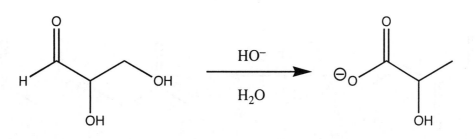

Chapter 24 Worksheet C

Provide a DETAILED, ARROW-PUSHING MECHANISM for the transformation below (16 pts).

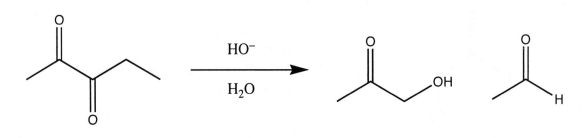

Name:_____   Date:_____

Chapter 24 Worksheet D

Provide a DETAILED, ARROW-PUSHING MECHANISM for the transformation below
(20 pts).

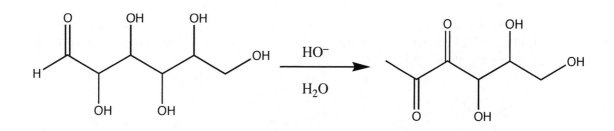

Chapter 24 Worksheet E

Many years ago, a very talented undergraduate assistant in my lab synthesized the unnatural sugar below by isolating an ALDOLASE enzyme from the key component of Bird's Nest Soup (available in some Chinese restaurants). One of the two components of this aldol reaction is pyruvate, and the other is a hexose derivative. DEDUCE THE STARTING MATERIAL FOR THIS ALDOL REACTION (8 pts). Draw the hexose STARTING MATERIAL in its hemiacetal form, in a chair conformation, with proper stereochemistry at any chiral centers (3 pts).

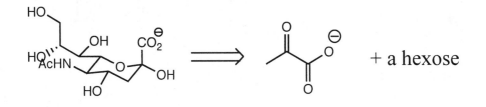

Name:_____   Date:_____

Chapter 25 Worksheet A

Circle the aromatic compounds from among the structures below (12 pts).

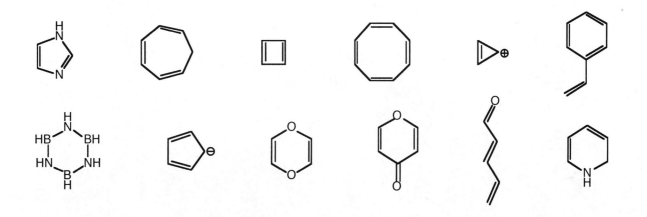

Chapter 25 Worksheet B

Imidazole has several lone pairs, having distinct basicity. Draw a three-dimensional representation of imidazole using the template provided below, including the specific orientation in space of each lone pair orbital (6 pts). Draw "x's" through electrons to count the number of π-electrons in a continuous ring in imidazole, to determine if it is aromatic (6 pts).

imidazole

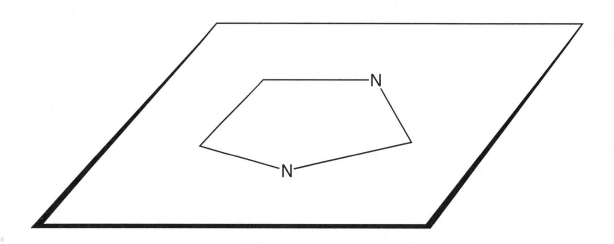

Chapter 25 Worksheet C

Professor K. Barry Sharpless (Nobel Laureate) once drew the molecule below on some scratch paper in my office, and excitedly suggested that it does "really neat chemistry." DRAW A DETAILED, ARROW-PUSHING MECHANISM for a reaction that this molecule can do that most alkenes would not dream of doing, USING ANY ADDITIONAL REAGENT (4 pts). Why does this reaction work (4 pts)?

an unusually stable compound

Chapter 25 Worksheet D

Addition of E⁺ to the heterocycle below occurs specifically at one carbon. Show a DETAILED, ARROW-PUSHING MECHANISM for formation of each possible INTERMEDIATE on addition of E⁺ to this heterocycle (9 pts). Draw the BEST RESONANCE FORM of the MOST STABLE INTERMEDIATE (2 pts), and complete the mechanism to afford the major product (3 pts).

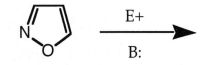

Chapter 25 Worksheet E

In the Pictet-Spengler reaction, an indole derivative bearing an amine side chain is treated with an aldehyde, to afford a larger aromatic heterocycle. Provide a DETAILED, ARROW-PUSHING MECHANISM for the Pictet-Spengler reaction below (12 pts).

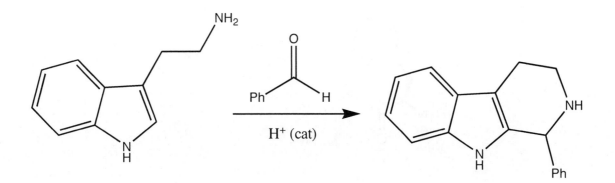

Chapter 26 Worksheet A

a) Identify the chiral center in PHENYLALANINE, with an asterisk next to that carbon (1 pt). Assign priorities to the non-hydrogen substituents (*a*, *b*, and *c*; 3 pts). Label the chiral center *R* or *S* (2 pts).

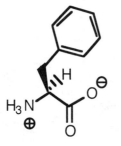

b) Label each distinct proton in PHENYLALANINE (A, B, C, …) (4 pts).

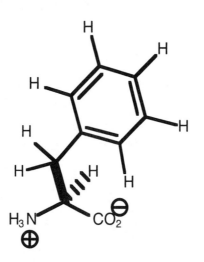

Chapter 26 Worksheet B

a) *Threonine*, like all amino acids, is both an acid and a base. The pK_a of the carboxylic acid moiety in protonated **II** below is 3.5, and the pK_a of the ammonium ion in **II** is 8.5. What is K_{eq} for the internal acid-base reaction shown (3 pts)? ESTIMATE $\Delta G°$ for this process (4 pts).

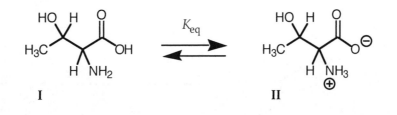

b) In AS FEW WORDS AS POSSIBLE, explain why the pK_as of *Threonine* are not the same as those of acetic acid and ammonium ion (3 pts).

Chapter 26 Worksheet C

Deduce the structure of the final product of the following reaction sequence (5 pts). The
^{1}H NMR data of the product, taken in $CDCl_3$ with a trace of D_2O, are provided. NOTE
THAT RAPIDLY EXCHANGING PROTONS WILL NOT BE OBSERVED IN THE
PRESENCE OF TRACE D_2O SOLVENT. Infer the connectivity of each distinct sp^3 C-H
proton from the ^{1}H NMR data below (5 pts). Label each one in your product (H$_A$, H$_B$,
etc., 5 pts).

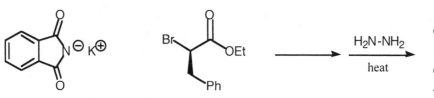

$C_{11}H_{15}NO_2$
^{1}H NMR (w/ trace D_2O)
6.5-7.5 ppm (m, 5H)
4.1 ppm (q, 2H) ==>
3.8 ppm (dd, 1H) ==>
3.2 ppm (Dd, 1H) ==>
3.1 ppm (Dd, 1H) ==>
1.3 ppm (t, 3H) ==>

Chapter 26 Worksheet D

Provide a DETAILED, ARROW-PUSHING MECHANISM for the reaction below
(20 pts).

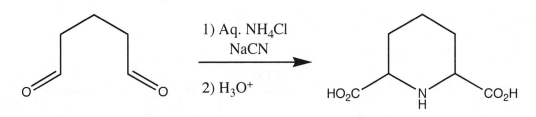

Chapter 26 Worksheet E

N-protected chiral amino acids sometimes lose their enantiomeric purity on treatment with reagents that activate the carbonyl group, via an intermediate that racemizes readily. Propose the structure of this intermediate formed from the amino acid derivative below (8 pts), and explain why it racemizes so quickly (4 pts).

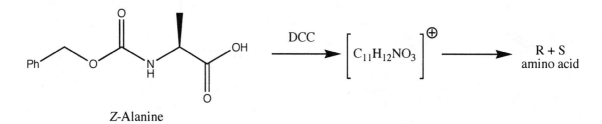

Z-Alanine